Biodiesel Breakthroughs Innovations and Advances in Renewable Energy

Sophia

TABLE OF CONTENTS

Chapter 1: Introduction to Biodiesel

Understanding Renewable Energy

Renewable energy is a term that has gained significant attention in recent years due to the growing concerns about climate change and the need for sustainable energy sources. In this subchapter, we will delve into the concept of renewable energy and its importance in addressing the global energy crisis. Specifically, we will focus on the niche of biofuels, particularly biodiesel, as a breakthrough innovation in the renewable energy sector.

Renewable energy refers to sources of energy that can be replenished naturally, such as solar, wind, hydro, and biofuels. Unlike fossil fuels, which are finite and contribute to greenhouse gas emissions, renewable energy provides a cleaner and more sustainable alternative. This subchapter aims to provide an accessible understanding of renewable energy for everyone, regardless of their background or expertise.

Within the realm of renewable energy, biofuels have emerged as a promising solution to reduce dependence on fossil fuels. Biofuels are derived from organic matter, commonly known as biomass, and can be used to power vehicles, machinery, and even generate electricity. Among various biofuels, biodiesel has gained considerable attention due to its compatibility with existing diesel engines and infrastructure.

Biodiesel breakthroughs have revolutionized the renewable energy landscape by utilizing waste oils, animal fats, and plant-based oils as feedstocks. The production process involves converting these raw materials into biodiesel through a chemical reaction called transesterification. Biodiesel not only reduces greenhouse gas emissions but also provides economic opportunities for farmers and promotes sustainable agriculture.

This subchapter will explore the advancements and innovations in biodiesel production, shedding light on the various feedstock options, production techniques, and their environmental impacts. Additionally, it will address the challenges and opportunities in the biofuel industry, including the development of efficient production methods, policies, and market integration.

Understanding renewable energy, particularly the niche of biofuels, is crucial for individuals, policymakers, and businesses alike. By embracing renewable energy sources like biodiesel, we can contribute to a cleaner and more sustainable future. This subchapter aims to empower readers with knowledge about renewable energy and inspire them to take action in their personal and professional lives.

In conclusion, this subchapter titled "Understanding Renewable Energy" provides an introductory overview of renewable energy, with a specific focus on biodiesel as a breakthrough in the biofuel niche. It aims to appeal to a wide audience, catering to individuals interested in renewable energy and its environmental benefits. By providing a comprehensive understanding of renewable energy and biodiesel, this subchapter encourages readers to explore the possibilities of sustainable energy sources and contribute to a greener future.

Importance of Biodiesel as a Renewable Energy Source

The Importance of Biodiesel as a Renewable Energy Source

Introduction:

In recent years, the global focus on renewable energy sources has intensified, as the world grapples with the challenges posed by climate change. One particular area of interest that has gained significant attention is biodiesel, a renewable energy source derived from organic materials such as vegetable oils and animal fats. In this subchapter, we will explore the importance of biodiesel as a sustainable alternative to conventional fossil fuels.

Environmental Benefits:

One of the key advantages of biodiesel is its positive impact on the environment. Unlike conventional fossil fuels, biodiesel produces significantly lower levels of greenhouse gas emissions, reducing the overall carbon footprint. It is a cleaner-burning fuel that releases fewer particulate matter and toxic pollutants, resulting in improved air quality and reduced health risks for everyone. Biodiesel also plays a crucial role in reducing dependence on finite fossil fuel resources, offering a sustainable and long-term energy solution.

Economic Advantages:

Apart from its environmental benefits, biodiesel holds significant economic advantages. As a renewable energy source, biodiesel production creates new job opportunities, particularly in the agricultural sector. It stimulates rural economies by encouraging local farmers to grow oilseed crops, providing them with an additional source of income. Moreover, biodiesel production promotes energy independence by reducing reliance on foreign oil imports, thus stabilizing fuel prices and strengthening national economies.

Technological Innovations:

The development of biodiesel has also spurred technological advancements in the renewable energy sector. Researchers and scientists continuously strive to improve the production process, making it more efficient and cost-effective. Breakthroughs in catalysts, feedstock development, and conversion technologies have enhanced biodiesel's viability as a mainstream fuel alternative. These innovations have paved the way for the integration of biodiesel in existing infrastructure and transportation systems, ensuring a smooth transition towards a sustainable energy future.

Conclusion:

In conclusion, biodiesel has emerged as a crucial renewable energy source with diverse benefits. Its environmental advantages, including reduced greenhouse gas emissions and improved air quality, make it a key player in mitigating climate change. The economic benefits, such as job creation and energy independence, contribute to a more sustainable and resilient economy. Furthermore, technological advancements have made biodiesel production more efficient and accessible, facilitating its integration into existing infrastructure. As we strive to address the global energy challenges, biodiesel stands as a promising solution, offering a cleaner, greener, and more sustainable energy future for everyone.

Historical Overview of Biodiesel Production

Biodiesel Breakthroughs: Innovations and Advances in Renewable Energy

Chapter X: Historical Overview of Biodiesel Production

Introduction:

In this subchapter, we will take a journey through the historical development of biodiesel production. Biodiesel is a renewable energy source derived from organic materials such as vegetable oils, animal fats, and even recycled cooking oil. It has gained significant attention in recent years as a viable alternative to fossil fuels due to its environmental benefits and potential to reduce greenhouse gas emissions. This chapter aims to provide an overview of the key milestones and advancements in biodiesel production, tracing its origins from ancient civilizations to the modern era.

Ancient Origins:

The use of vegetable oils and fats as fuel can be traced back to ancient civilizations such as the Egyptians, who utilized oilseeds to produce fuel for lamps. Similarly, the Greeks and Romans utilized olive oil as a source of lighting fuel. These early discoveries laid the foundation for the later development of biodiesel production techniques.

Industrial Revolution:

The Industrial Revolution marked a turning point in the history of energy production. With the advent of steam engines, the demand for fossil fuels skyrocketed. However, at the same time, researchers began exploring alternative fuels. In the 19th century, Rudolf Diesel, the inventor of the diesel engine, experimented with vegetable oils and successfully ran his engine on peanut oil. This breakthrough laid the groundwork for the future development of biodiesel.

Modern Era:

The modern era witnessed a resurgence of interest in biodiesel production due to concerns over climate change and the need for sustainable energy sources. In the 1970s, during the oil crisis, biodiesel gained attention as a potential solution to reduce dependence on fossil fuels. Research and development efforts intensified, leading to the establishment of the first biodiesel production plants in the 1990s.

Advancements and Innovations:

Over the years, significant advancements have been made in biodiesel production. Researchers have focused on improving conversion efficiency and reducing production costs. One notable breakthrough was the development of transesterification, a chemical process that converts vegetable oils or animal fats into biodiesel. This technique revolutionized the industry by streamlining the production process and making biodiesel more commercially viable.

Conclusion:

The historical overview of biodiesel production highlights the long-standing human fascination with finding sustainable energy alternatives. From ancient civilizations to the modern era, biodiesel has evolved from a simple lighting fuel to a promising renewable energy source. The continuous advancements and innovations in biodiesel production techniques hold great potential for a greener and more sustainable future.

Chapter 2: Basics of Biodiesel Production

Feedstock Selection for Biodiesel Production

Biodiesel has emerged as a promising alternative to conventional fossil fuels, offering numerous environmental and economic benefits. However, the selection of appropriate feedstock is crucial for the successful production of biodiesel. This subchapter aims to provide an overview of the various factors that need to be considered when selecting feedstock for biodiesel production.

The choice of feedstock for biodiesel production depends on several factors, including availability, cost, sustainability, and the quality of the resulting biodiesel. The primary feedstocks used in biodiesel production include vegetable oils, animal fats, and waste oils. Each feedstock has its own unique characteristics, which influence its suitability for biodiesel production.

Vegetable oils, such as soybean, canola, and palm oil, are the most commonly used feedstocks for biodiesel production. They have high oil content and can be easily converted into biodiesel through a process called transesterification. However, the use of vegetable oils as feedstock has raised concerns over deforestation and land-use change.

Animal fats, obtained from livestock and poultry processing industries, are another potential feedstock for biodiesel production. These fats are considered as waste products and can be a cost-effective option for biodiesel production. However, animal fat feedstocks may have higher levels of impurities, requiring additional processing steps.

Waste oils, such as used cooking oils and trap greases, are a sustainable feedstock option for biodiesel production. These oils are typically collected from restaurants and food processing industries, reducing the environmental impact of waste disposal. However, waste oils may contain higher levels of impurities and require thorough pretreatment before being used for biodiesel production.

The selection of feedstock also depends on the climate and agricultural practices of a region. For example, regions with a suitable climate for growing oilseed crops may prefer vegetable oils as feedstock. Conversely, areas with a significant livestock industry may opt for animal fats as a feedstock.

In conclusion, the selection of feedstock plays a crucial role in the production of biodiesel. Factors such as availability, cost, sustainability, and feedstock characteristics need to be carefully considered. Vegetable oils, animal fats, and waste oils are the primary feedstock options, each with its own advantages and considerations. By making informed choices, we can ensure the production of biodiesel that is sustainable, cost-effective, and environmentally friendly.

This subchapter aims to provide a comprehensive understanding of the feedstock selection process, enabling readers interested in biofuels to make informed decisions and contribute to the advancement of renewable energy.

Transesterification Process

In the realm of renewable energy, biofuels have captured the attention of scientists, policymakers, and consumers alike. The quest for sustainable alternatives to traditional fossil fuels has led to significant breakthroughs in the field, and one such innovation is the transesterification process. This subchapter aims to provide an accessible overview of this crucial step in biodiesel production, making it easily understandable for everyone interested in biofuels.

Transesterification is a chemical reaction that converts vegetable oils or animal fats into biodiesel by replacing the glycerin molecule in the oil with an alcohol molecule, typically methanol or ethanol. This process can be catalyzed by either an acid or a base, but the most commonly used catalyst is sodium hydroxide or potassium hydroxide, known as alkaline transesterification. The reaction takes place in the presence of heat, and the resulting product is a cleaner-burning fuel that can be used in existing diesel engines without any modifications.

The transesterification process offers several advantages over other methods of biodiesel production. Firstly, it is a relatively simple and cost-effective process, making it commercially viable for large-scale production. Additionally, it can utilize a wide range of feedstocks, including waste cooking oil, algae, and even used animal fats, reducing the environmental impact of these waste products.

The subchapter will delve into the intricacies of the transesterification process, explaining the role of catalysts, temperature, and pressure in optimizing the reaction. It will also explore the various factors that can affect the efficiency of the process, such as the quality of feedstock and the presence of impurities. Real-life examples and case studies will be provided to illustrate the practical applications of transesterification in the biofuel industry.

Furthermore, the subchapter will touch upon the ongoing research and advancements in transesterification technology, including the development of heterogenous catalysts, enzymatic transesterification, and microwave-assisted transesterification. These innovations have the potential to further enhance the efficiency, sustainability, and scalability of biodiesel production.

In conclusion, the transesterification process is a crucial step in the production of biodiesel, offering a sustainable alternative to traditional fossil fuels. This subchapter aims to provide a comprehensive yet accessible overview of the process, catering to the interests of both biofuel enthusiasts and those seeking a deeper understanding of renewable energy. By highlighting the advantages, challenges, and ongoing advancements in transesterification, it seeks to inspire further exploration and adoption of biofuels as a viable solution for a greener future.

Catalysts Used in Biodiesel Production

Biodiesel has emerged as a promising alternative to fossil fuels, offering a more sustainable and environmentally friendly energy source. One of the key components in the production of biodiesel is the catalyst, which plays a crucial role in converting vegetable oils or animal fats into biodiesel through a process called transesterification.

Transesterification involves the reaction of triglycerides with an alcohol, typically methanol, in the presence of a catalyst. The catalyst facilitates the reaction by increasing the rate of conversion and enhancing the efficiency of the process. There are various catalysts used in biodiesel production, each with its own advantages and limitations.

One commonly used catalyst is sodium hydroxide or potassium hydroxide, also known as alkaline catalysts. These catalysts are highly effective and cost-efficient, making them popular choices in industrial-scale biodiesel production. They promote rapid reaction kinetics and allow for high yields of biodiesel. However, they have some drawbacks, such as the requirement for careful handling due to their corrosive nature and the generation of significant amounts of waste by-products.

Another type of catalyst used in biodiesel production is acid catalysts, such as sulfuric acid or hydrochloric acid. Acid catalysts are often employed in small-scale or laboratory settings due to their lower cost and ease of use. However, they have slower reaction rates compared to alkaline catalysts and may require additional steps to neutralize the acidic by-products.

Enzymes are a more recent development in biodiesel catalysts. These biocatalysts offer several advantages, including high specificity and lower energy requirements. Enzymes can operate at lower temperatures and pressures, reducing the energy input and overall production costs. However, they are still in the early stages of commercialization, and further research is needed to optimize their performance and reduce their production costs.

In recent years, heterogeneous catalysts have gained attention for their potential in biodiesel production. These catalysts are solid materials, such as zeolites or metal oxides, which can be reused and offer the advantage of easy separation from the reaction mixture. Heterogeneous catalysts have shown promise in improving reaction rates and reducing waste generation.

The choice of catalyst depends on several factors, including the scale of production, cost considerations, reaction conditions, and desired biodiesel quality. Ongoing research and development efforts are focused on finding more efficient and sustainable catalysts to enhance the overall biodiesel production process.

In conclusion, catalysts play a vital role in the conversion of vegetable oils or animal fats into biodiesel. The choice of catalyst influences the overall efficiency, cost-effectiveness, and sustainability of the biodiesel production process. Scientists and researchers continue to explore and innovate in this field, aiming to develop more efficient and environmentally friendly catalysts for a greener energy future.

Chapter 3: Advances in Feedstock for Biodiesel Production

Traditional Feedstock Sources (Soybean, Rapeseed, etc.)

The use of traditional feedstock sources such as soybean, rapeseed, and others has played a significant role in the development and production of biodiesel. These feedstocks have been widely utilized due to their abundance, availability, and favorable oil composition. In this subchapter, we will explore the various traditional feedstock sources used in biodiesel production and their importance in the biofuels industry.

Soybean, a versatile legume, has been a popular feedstock for biodiesel production. It is widely cultivated in several countries, making it easily accessible and cost-effective. Soybean oil has a high energy content and favorable fatty acid profile, making it an ideal choice for biodiesel production. Additionally, soybean cultivation provides an opportunity for crop rotation, which enhances soil fertility and reduces the need for synthetic fertilizers.

Rapeseed, also known as canola, is another important traditional feedstock for biodiesel. It is widely grown worldwide and possesses a high oil content. Rapeseed oil is low in saturated fat and high in monounsaturated and polyunsaturated fats, making it a healthier alternative to conventional diesel fuel. The cultivation of rapeseed also contributes to sustainable agriculture practices, as it requires fewer pesticides and herbicides compared to other crops.

Other traditional feedstock sources used in biodiesel production include sunflower, palm, and jatropha. Sunflower oil is rich in vitamin E and has similar properties to soybean oil, making it a suitable alternative. Palm oil, although controversial due to its impact on deforestation and biodiversity, remains a significant feedstock due to its high yield and favorable oil composition. Jatropha, a non-edible oilseed plant, is gaining popularity as a biodiesel feedstock due to its ability to grow in arid regions with minimal water requirements.

In conclusion, traditional feedstock sources such as soybean, rapeseed, sunflower, palm, and jatropha have been instrumental in the development and advancement of biodiesel production. Their abundance, oil composition, and potential for sustainable cultivation make them attractive options for the biofuels industry. As the demand for renewable energy continues to rise, it is crucial to explore and optimize the utilization of these traditional feedstocks to ensure a sustainable and environmentally friendly future for biofuels.

Note: The content provided is for informational purposes only and does not endorse or promote the use of any specific feedstock. The impact of feedstock selection on sustainability and the environment should be thoroughly assessed and considered.

Algae as a Promising Feedstock for Biodiesel Production

In recent years, the search for renewable energy sources has intensified, and biodiesel has emerged as a promising alternative to conventional fossil fuels. Biodiesel is a renewable, non-toxic, and biodegradable fuel that can be produced from a variety of feedstocks, including vegetable oils, animal fats, and even algae.

Algae, often disregarded as a mere nuisance, have gained significant attention as a potential feedstock for biodiesel production. This subchapter explores the reasons behind algae's rising popularity and its immense potential in shaping the future of biofuels.

One of the key advantages of algae lies in its high lipid content, which can be converted into biodiesel through a process called transesterification. Compared to traditional feedstocks like soybeans or rapeseed, algae can produce up to 100 times more oil per unit of land area, making it a highly efficient option for biodiesel production. Additionally, algae can be grown in various environments, including wastewater, saltwater, or even in specially designed photobioreactors, reducing the pressure on valuable agricultural land.

Furthermore, algae have a rapid growth rate, with some species doubling in biomass within a single day. This fast growth enables continuous production of biodiesel throughout the year, unlike traditional crops that have limited growing seasons. The ability to cultivate algae year-round ensures a stable and reliable source of biodiesel.

Algae cultivation also offers environmental benefits. Algae can consume carbon dioxide (CO_2) during photosynthesis, acting as a natural carbon sink and reducing greenhouse gas emissions. By utilizing algae for biodiesel production, we can simultaneously reduce CO2 levels in the atmosphere and produce a clean, sustainable fuel.

Moreover, the byproducts of algae cultivation can have additional value. Algae biomass can be used as a nutrient-rich fertilizer or as feed for livestock, further enhancing the economic feasibility of algae-based biodiesel production.

Despite its numerous advantages, there are still challenges to overcome in scaling up algae cultivation for commercial biodiesel production. These include optimizing algae growth conditions, improving extraction techniques, and developing cost-effective harvesting methods. However, ongoing research and technological advancements are steadily addressing these challenges, bringing us closer to a future where algae-based biodiesel becomes a mainstream renewable energy source.

In conclusion, algae represent a highly promising feedstock for biodiesel production. With its high oil content, rapid growth rate, and ability to thrive in diverse environments, algae offer a sustainable and efficient solution to the growing demand for renewable fuels. By harnessing the potential of algae, we can significantly reduce our reliance on fossil fuels and pave the way for a greener, cleaner future for everyone.

Waste Cooking Oil as a Sustainable Feedstock

In recent years, there has been a growing interest in finding sustainable alternatives to traditional fossil fuels. One such alternative that has gained significant attention in the field of biofuels is waste cooking oil. This renewable energy source holds immense promise as a sustainable feedstock for biodiesel production.

Every household generates waste cooking oil, which is usually discarded as a problem rather than considered as a valuable resource. However, this waste can be transformed into a sustainable feedstock for the production of biodiesel, a clean-burning and renewable fuel. The use of waste cooking oil as a feedstock not only helps in reducing environmental pollution but also provides a cost-effective solution to the ever-increasing demand for energy.

Biofuels represent a niche in the renewable energy sector that has the potential to significantly reduce greenhouse gas emissions and dependence on fossil fuels. Biodiesel, a type of biofuel derived from organic sources, offers a sustainable alternative to petroleum-based diesel. By utilizing waste cooking oil as a feedstock for biodiesel production, we can reduce the carbon footprint associated with the transportation sector and contribute to a greener future.

The advantages of using waste cooking oil as a feedstock for biodiesel production are manifold. Firstly, it offers a way to repurpose a waste product, reducing the strain on landfill sites and preventing pollution caused by improper disposal. Additionally, waste cooking oil is readily available and can be easily collected from households, restaurants, and food processing industries. This accessibility makes it a cost-effective feedstock option, especially when compared to other biofuel sources that require extensive agricultural land or complex extraction processes.

Furthermore, the use of waste cooking oil as a feedstock contributes to the circular economy concept. By converting waste into a valuable resource, we can create a closed-loop system where waste is recycled and repurposed rather than discarded. This not only reduces the environmental impact but also promotes sustainability and resource efficiency.

In conclusion, waste cooking oil has emerged as a sustainable feedstock for biodiesel production. By utilizing this readily available waste product, we can reduce pollution, decrease dependence on fossil fuels, and contribute to a greener future. It is crucial that we recognize the potential of waste cooking oil and work towards implementing sustainable practices that harness its energy potential. By doing so, we can make significant strides in advancing the field of biofuels and creating a more sustainable and environmentally friendly energy sector.

Chapter 4: Innovations in Biodiesel Production Technology

Enzymatic Transesterification Process

In recent years, there has been a growing interest in finding sustainable and renewable sources of energy, and biofuels have emerged as a promising solution. One of the key advancements in the field of biofuels is the enzymatic transesterification process. This innovative technique has revolutionized the production of biodiesel, making it more efficient, environmentally friendly, and economically viable.

Traditionally, biodiesel was produced through a chemical process called transesterification, which involved the reaction of vegetable oils or animal fats with an alcohol, typically methanol, in the presence of a catalyst such as sodium hydroxide. However, this method had its limitations, including the need for high temperatures, long reaction times, and the production of significant amounts of waste.

The enzymatic transesterification process, on the other hand, eliminates many of these drawbacks. It utilizes specific enzymes, known as lipases, as the catalysts for the reaction. Lipases are naturally occurring biocatalysts that can accelerate the reaction between the oil or fat and the alcohol, thereby facilitating the conversion into biodiesel.

This enzymatic process offers several advantages over the conventional chemical method. Firstly, it operates at lower temperatures and atmospheric pressure, reducing energy consumption and production costs. Secondly, it has a shorter reaction time, resulting in higher productivity. Additionally, the enzymatic transesterification process generates less waste and has a lower environmental impact, making it more sustainable and in line with the principles of green chemistry.

Moreover, this process can utilize a wide range of feedstocks, including low-quality oils, waste cooking oils, and even algae, further expanding the potential sources of biodiesel production. It also produces a higher purity biodiesel, with fewer impurities, enhancing the fuel quality and reducing engine emissions.

The enzymatic transesterification process has been widely researched and implemented in various biodiesel production plants across the globe. Its scalability and compatibility with existing infrastructure make it a viable option for large-scale biofuel production. Furthermore, ongoing research continues to improve the efficiency and cost-effectiveness of this process, paving the way for a more sustainable future.

In conclusion, the enzymatic transesterification process is a breakthrough innovation in the field of biofuels. It offers a more efficient, eco-friendly, and economically viable alternative to traditional chemical transesterification methods. With its numerous advantages and potential for scalability, this process has the potential to revolutionize the biofuel industry, contributing to a cleaner and more sustainable energy future for everyone.

Supercritical Methanol Process

The Supercritical Methanol Process is a groundbreaking innovation in the field of biofuels, specifically biodiesel production. This process holds immense potential for revolutionizing the renewable energy sector by providing a more efficient and sustainable method of converting feedstocks into high-quality biodiesel.

In this subchapter, we will delve into the intricacies of the Supercritical Methanol Process, shedding light on its various aspects, advantages, and implications. This content aims to educate and engage a broad audience, from individuals interested in renewable energy to experts in the field of biofuels.

The Supercritical Methanol Process involves the use of supercritical methanol, a state wherein methanol is heated above its critical temperature and pressure, resulting in a unique solvent with superior properties. This supercritical methanol acts as a catalyst, significantly enhancing the transesterification reaction that converts feedstocks, such as vegetable oils or animal fats, into biodiesel.

One of the key advantages of the Supercritical Methanol Process is its ability to eliminate the need for additional catalysts, thus simplifying the production process and reducing costs. Moreover, the supercritical methanol offers higher conversion efficiencies and faster reaction rates, leading to increased biodiesel yields compared to traditional methods.

Another significant benefit of this process is its ability to utilize a wide range of feedstocks, including low-quality or waste oils, thereby reducing dependence on food crops and minimizing environmental impact. By converting these feedstocks into biodiesel, the Supercritical

Methanol Process contributes to waste reduction and the promotion of a circular economy.

Furthermore, the use of supercritical methanol in biodiesel production results in cleaner and more sustainable fuel. Biodiesel produced through this process exhibits superior fuel properties, such as improved cold flow, lower viscosity, and reduced emissions of harmful pollutants. This makes it an attractive alternative to conventional diesel, helping to mitigate climate change and improve air quality.

Overall, the Supercritical Methanol Process represents a significant breakthrough in the field of biofuels. Its potential to increase efficiency, simplify production, and utilize various feedstocks makes it a promising solution for achieving a more sustainable and renewable energy future. By understanding and embracing this innovative process, we can accelerate the transition towards a cleaner and greener world.

Microwave-Assisted Production of Biodiesel

In recent years, the search for alternative energy sources has gained significant momentum. As the world grapples with the consequences of climate change and the depletion of traditional fossil fuels, the need for sustainable and renewable energy solutions has become more pressing than ever. Among the many breakthroughs in this field, microwave-assisted production of biodiesel has emerged as a promising innovation.

Biodiesel, derived from renewable sources such as vegetable oils and animal fats, has gained popularity as a cleaner and more environmentally friendly alternative to traditional diesel fuel. However, the conventional production methods for biodiesel are time-consuming and often require high temperatures and large amounts of chemicals. This is where microwave-assisted production comes into play.

Microwave-assisted production of biodiesel involves the use of microwaves to accelerate the chemical reactions involved in the conversion of oils and fats into biodiesel. This method offers several advantages over traditional techniques. Firstly, it significantly reduces the reaction time, allowing for a faster production process. Secondly, it requires lower temperatures, reducing energy consumption and minimizing the environmental impact. Lastly, microwave-assisted production eliminates the need for excessive chemicals, making it a more sustainable and cost-effective option.

The application of microwaves in biodiesel production has shown promising results. Studies have demonstrated improved yield and quality of biodiesel using this method, with shorter reaction times and higher conversion rates. Furthermore, microwave-assisted production allows for better control over the reaction conditions, leading to more consistent and reproducible results.

While microwave-assisted production of biodiesel is still a relatively new concept, its potential benefits make it an area of great interest for researchers and industry professionals. The utilization of microwaves in the process can revolutionize the biodiesel industry by improving efficiency, reducing costs, and minimizing the environmental footprint.

In conclusion, microwave-assisted production of biodiesel holds great promise in the quest for sustainable energy solutions. With its ability to accelerate the conversion process, reduce energy consumption, and minimize chemical usage, this innovation has the potential to make biodiesel production more efficient and environmentally friendly. As researchers continue to explore and refine this technology, microwave-assisted production could become a key player in the biofuels industry, contributing to a greener and more sustainable future for all.

Chapter 5: Improving Biodiesel Quality and Performance

Catalysts for Enhanced Biodiesel Yield

In recent years, biodiesel has emerged as a promising alternative to fossil fuels, offering a sustainable and environmentally friendly source of energy. As the demand for renewable energy continues to grow, researchers and scientists have been striving to enhance the production of biodiesel. One key aspect that has played a significant role in this pursuit is the development of catalysts that can improve the biodiesel yield.

Catalysts act as facilitators in chemical reactions, speeding up the conversion of raw materials into biodiesel. They play a crucial role in the transesterification process, where vegetable oils or animal fats are converted into biodiesel by reacting with an alcohol, such as methanol or ethanol. By using catalysts, this reaction can occur at a faster rate and under milder conditions, resulting in higher biodiesel yields.

Several catalysts have been investigated and developed to enhance biodiesel production. One of the most commonly used catalysts is sodium or potassium hydroxide, which is an alkali catalyst. Alkali catalysts exhibit high conversion rates and are relatively cost-effective, making them widely adopted in the industry. However, they also have drawbacks, such as the formation of soap as a byproduct and the requirement for high-quality feedstock.

To overcome these limitations, acid catalysts, such as sulfuric acid and hydrochloric acid, have been explored. Acid catalysts can convert low-quality feedstock into biodiesel efficiently, reducing the dependency on high-quality oils or fats. Additionally, they produce lower amounts of soap during the reaction, simplifying the purification process.

Another category of catalysts that has gained attention is enzyme catalysts. Enzymes are natural catalysts produced by living organisms, such as bacteria or fungi. They offer several advantages, including high specificity, mild reaction conditions, and reduced environmental impact. However, enzyme catalysts are often expensive and require careful handling.

In recent years, researchers have also focused on developing heterogeneous catalysts, which are solid materials that facilitate the transesterification reaction. These catalysts can be reused multiple times and offer better control over the reaction conditions.

In conclusion, catalysts play a crucial role in enhancing biodiesel yield by accelerating the transesterification process. Various catalysts, including alkali, acid, enzyme, and heterogeneous catalysts, have been investigated for their ability to improve the efficiency and sustainability of biodiesel production. The choice of catalyst depends on factors such as feedstock quality, cost-effectiveness, and environmental impact. Continued research and innovation in catalyst development are essential to further optimize biodiesel production and pave the way for a greener future.

Blending Biodiesel with Conventional Diesel (Biodiesel-Diesel Blends)

In recent years, the search for renewable energy sources has gained significant traction. One such promising alternative to traditional fossil fuels is biodiesel. Derived from natural sources such as vegetable oils, animal fats, and recycled cooking oil, biodiesel offers numerous advantages over conventional diesel. However, the transition to biodiesel on a large scale is not without its challenges.

Blending biodiesel with conventional diesel is a practical and effective approach to overcome some of these challenges. Biodiesel-diesel blends combine a certain percentage of biodiesel with regular diesel fuel, creating a product that can be used in existing diesel engines without any significant modifications. This blending process allows for a gradual integration of biodiesel into the existing infrastructure, making it accessible to a wider audience.

One of the key benefits of biodiesel-diesel blends is their reduced environmental impact. Biodiesel is a cleaner-burning fuel that produces lower levels of greenhouse gas emissions compared to traditional diesel. By blending biodiesel with conventional diesel, the overall emissions from diesel engines can be significantly reduced, contributing to a healthier and more sustainable environment.

Another advantage of biodiesel-diesel blends is their compatibility with existing engines and infrastructure. Unlike other alternative fuels, such as hydrogen or electric power, biodiesel-diesel blends can be used in conventional diesel engines without the need for extensive modifications or infrastructure investments. This makes the transition to biodiesel more feasible and cost-effective for a wide range of applications, from transportation to stationary power generation.

Furthermore, biodiesel-diesel blends offer improved lubricity, which helps protect and extend the life of engine components. The higher lubricity provided by biodiesel reduces wear and tear on engine parts, resulting in reduced maintenance costs and increased engine durability.

It is worth noting that the blending ratio of biodiesel and diesel can vary depending on factors such as climate, engine type, and emission regulations. Common blends include B5 (5% biodiesel, 95% diesel), B20 (20% biodiesel, 80% diesel), and B100 (pure biodiesel). The choice of blend depends on the specific requirements of the application and the availability of biodiesel in the region.

In conclusion, blending biodiesel with conventional diesel offers a practical and accessible solution for transitioning to renewable energy sources. Biodiesel-diesel blends provide reduced emissions, improved lubricity, and compatibility with existing engines and infrastructure. As the demand for sustainable energy grows, biodiesel-diesel blends have the potential to play a significant role in reducing our dependence on fossil fuels and mitigating the environmental impact of transportation and power generation.

Additives for Improved Cold Flow Properties and Stability

In the ever-evolving world of renewable energy, biodiesel has emerged as a promising alternative to traditional fossil fuels. Derived from renewable sources such as vegetable oils and animal fats, biodiesel offers several environmental benefits, including reduced greenhouse gas emissions and decreased dependence on finite resources. However, like any other fuel, biodiesel faces certain challenges that need to be addressed for its widespread adoption. One such challenge is the cold flow properties and stability of biodiesel.

Biodiesel, being composed of long-chain fatty acid methyl esters, tends to solidify at low temperatures, resulting in poor cold flow properties. This can lead to clogged fuel filters, reduced engine performance, and even engine failure in extreme cases. To overcome this limitation, researchers and scientists have developed various additives that improve the cold flow properties of biodiesel, making it more suitable for use in colder climates.

One of the most commonly used additives is pour point depressants (PPDs). These additives modify the crystallization behavior of biodiesel, preventing the formation of large, solid crystals that can obstruct fuel flow. PPDs work by interfering with the crystal growth process, reducing the pour point temperature and improving the cold flow properties of biodiesel.

Another crucial aspect of biodiesel stability is oxidative stability, which refers to its resistance to degradation caused by exposure to oxygen. Oxidation leads to the formation of harmful compounds that can negatively impact fuel quality and engine performance. To enhance the oxidative stability of biodiesel, antioxidants are added. These additives inhibit the oxidation process, ensuring that biodiesel maintains its integrity and performance over time.

Furthermore, stability during storage and transportation is of utmost importance to ensure the quality of biodiesel. Antioxidants also play a significant role in improving the stability of biodiesel by preventing the formation of sediments and gums that can clog fuel lines and injectors.

In conclusion, the development of additives for improved cold flow properties and stability is crucial for the advancement and wider acceptance of biodiesel as a viable alternative to conventional fuels. By addressing the challenges related to cold weather performance and stability, these additives ensure that biodiesel can be used reliably in all climates and conditions. With ongoing research and innovation in this field, biodiesel continues to make significant breakthroughs, contributing to a cleaner and more sustainable future for everyone interested in the niche of biofuels.

Chapter 6: Environmental and Economic Impacts of Biodiesel Production

Reducing Greenhouse Gas Emissions through Biodiesel Use

Biodiesel Breakthroughs: Innovations and Advances in Renewable Energy

Introduction:

In recent years, there has been a growing concern about the negative impact of greenhouse gas emissions on our environment and the urgent need to transition to cleaner and more sustainable energy sources. Biodiesel, derived from renewable resources such as vegetable oils and animal fats, has emerged as a promising alternative to conventional fossil fuels. This subchapter aims to explore the potential of biodiesel in reducing greenhouse gas emissions and its significance in the broader context of renewable energy.

Understanding Greenhouse Gas Emissions:

Before delving into the benefits of biodiesel, it is crucial to understand the concept of greenhouse gas emissions. These gases, including carbon dioxide (CO_2), methane (CH_4), and nitrous oxide (N_2O), trap heat in the Earth's atmosphere, leading to global warming and climate change. The burning of fossil fuels, such as gasoline and diesel, is a major contributor to these emissions.

The Role of Biodiesel:

Biodiesel offers a sustainable and eco-friendly solution to combat greenhouse gas emissions. When compared to traditional petroleum-based diesel, biodiesel significantly reduces CO_2 emissions. This reduction is attributed to the fact that the carbon released during biodiesel combustion is offset by the carbon absorbed by the plants used to produce the fuel. Biodiesel also produces fewer harmful pollutants, such as sulfur and particulate matter, resulting in improved air quality.

Environmental Benefits:

The use of biodiesel contributes to several environmental benefits. Firstly, it aids in reducing our dependence on finite fossil fuel reserves, promoting energy security. Secondly, biodiesel production creates opportunities for recycling waste oils and fats, reducing landfill waste and pollution. Additionally, biodiesel production plants can contribute to the creation of jobs in rural areas, boosting local economies.

Challenges and Future Prospects:

While biodiesel presents numerous advantages, challenges remain in its widespread adoption. These include the availability and cost of feedstock, compatibility with existing diesel engines, and the need for appropriate infrastructure for production and distribution. However, ongoing research and technological advancements are addressing these challenges, making biodiesel a viable and attractive option for reducing greenhouse gas emissions.

Conclusion:

Biodiesel holds immense potential in reducing greenhouse gas emissions and mitigating the adverse effects of climate change. Its renewable nature, reduced carbon footprint, and multiple environmental benefits make it an essential component of the transition towards a more sustainable energy future. By embracing biodiesel, we can collectively contribute to a cleaner and healthier planet for future generations.

Life Cycle Assessment of Biodiesel Production

Life Cycle Assessment (LCA) is a crucial tool used to evaluate the environmental impact of any product or process. In the case of biodiesel production, LCA plays a vital role in assessing its sustainability and identifying areas for improvement. This subchapter will delve into the Life Cycle Assessment of Biodiesel Production, shedding light on its significance and providing insights into the environmental performance of biodiesel as a renewable energy source.

Life Cycle Assessment is a comprehensive approach that considers the entire life cycle of a product or process, from extraction of raw materials through production, use, and disposal. For biodiesel, this entails analyzing each stage involved in the production process, including feedstock cultivation, transportation, conversion, and distribution.

During the cultivation phase, LCA evaluates the environmental impacts associated with growing feedstock crops, such as soybeans, rapeseed, or palm oil. Factors such as land use, water consumption, and the use of fertilizers and pesticides are assessed to determine the sustainability of these crops. LCA also analyzes the transportation of feedstock to the biodiesel production facility, taking into account energy consumption and emissions associated with transportation methods.

The conversion phase focuses on the production of biodiesel from feedstock. LCA assesses the energy inputs required for the conversion process, such as heating, mixing, and separation. It also considers the emissions generated during these processes, including greenhouse gas emissions, air pollutants, and wastewater generation.

The distribution phase examines the transportation of biodiesel to end-users, analyzing the energy consumption and emissions associated with storage, loading, and transportation methods. Additionally, LCA evaluates the environmental impact of using biodiesel as a fuel, including exhaust emissions and potential impacts on air quality.

By conducting a Life Cycle Assessment, stakeholders in the biodiesel industry can gain a comprehensive understanding of the environmental impacts of biodiesel production. This assessment helps identify areas for improvement and supports the implementation of sustainable practices. For instance, LCA can highlight the benefits of using certain feedstocks or production methods that have lower environmental impacts.

Overall, the Life Cycle Assessment of Biodiesel Production is a valuable tool for measuring the environmental performance of biodiesel as a renewable energy source. It provides a scientific basis for decision-making, enabling governments, policymakers, and industry professionals to make informed choices to minimize the environmental footprint of biodiesel production and ensure a sustainable future for the biofuels industry.

Economic Viability and Market Potential of Biodiesel

In recent years, renewable energy sources have gained significant attention as society seeks to reduce its dependence on fossil fuels and combat the negative environmental impacts associated with them. Among these alternative energy solutions, biodiesel has emerged as a promising option due to its numerous advantages and potential for economic viability. This subchapter explores the economic aspects and market potential of biodiesel, shedding light on its role in the biofuel industry.

Biodiesel, a renewable fuel derived from vegetable oils or animal fats, offers several benefits compared to traditional petroleum-based diesel. Firstly, it significantly reduces greenhouse gas emissions, addressing one of the major contributors to climate change. Additionally, biodiesel is biodegradable and non-toxic, making it more environmentally friendly and safer to handle than conventional diesel. These characteristics, coupled with the fact that it can be seamlessly integrated into existing diesel infrastructure, make biodiesel an attractive option for many industries and consumers.

From an economic perspective, biodiesel presents a compelling case. As fossil fuel prices continue to fluctuate and concerns over energy security rise, biodiesel offers a stable and renewable energy source. Its production can be localized, reducing dependence on foreign oil and creating job opportunities in the biofuel industry. Moreover, biodiesel production can stimulate rural economies by utilizing local agricultural resources and providing an additional income stream for farmers.

The market potential for biodiesel is vast, with both large-scale commercial applications and individual consumers driving demand.

Industries such as transportation, agriculture, and construction are increasingly adopting biodiesel to meet their energy needs while reducing their carbon footprint. Additionally, individual consumers are becoming more conscious of their environmental impact and are actively seeking sustainable alternatives. Biodiesel fills this gap, providing a clean and renewable fuel option for vehicles and home heating systems.

Furthermore, governments worldwide are implementing policies and regulations to promote the use of renewable energy sources, including biodiesel. Incentives such as tax credits, grants, and renewable fuel standards are encouraging the adoption of biodiesel and creating a favorable market environment. These initiatives not only contribute to the economic viability of biodiesel but also demonstrate a commitment to sustainable energy solutions.

In conclusion, biodiesel holds immense economic viability and market potential in the biofuel industry. Its environmental benefits, compatibility with existing infrastructure, and government support make it an attractive alternative to traditional diesel. As society continues to prioritize renewable energy sources, biodiesel breakthroughs and innovations will play a crucial role in shaping a sustainable and greener future for everyone.

Chapter 7: Challenges and Future Perspectives in Biodiesel Production

Overcoming Feedstock Availability and Cost Issues

In recent years, the demand for renewable energy sources has surged, and the biofuels industry has emerged as a promising solution to reduce our dependence on fossil fuels. However, one of the main challenges faced by the biofuels industry is the availability and cost of feedstock.

Feedstock, which refers to the raw materials used to produce biofuels, can vary widely depending on the region and the type of biofuel being produced. Common feedstocks include vegetable oils, animal fats, and waste materials such as used cooking oil and algae. While these feedstocks offer great potential, their availability and cost can pose significant obstacles to the widespread adoption of biofuels.

One of the key strategies for overcoming feedstock availability issues is diversification. By exploring a wide range of feedstock options, we can mitigate the risk associated with relying solely on a single source. For instance, instead of solely using vegetable oils, which can be limited and expensive, we can also consider utilizing waste materials that are abundant and often discarded. This approach not only reduces feedstock costs but also helps in waste management and promotes sustainability.

Furthermore, technological advancements have played a crucial role in expanding the range of viable feedstocks. Researchers and scientists are continuously working on developing new processes and techniques to convert various feedstocks into biofuels efficiently. For example, the use of genetically modified microorganisms can enhance the conversion efficiency of feedstocks like algae, which have a high potential for biodiesel production.

Another factor to consider is the development of sustainable farming practices. By promoting sustainable agriculture techniques, we can ensure a consistent supply of feedstock without depleting natural resources or causing harm to the environment. This can include practices such as crop rotation, organic farming methods, and responsible land management.

To address the issue of feedstock costs, partnerships and collaborations are vital. Governments, academia, and industry stakeholders must work together to find innovative solutions and create a favorable market environment for biofuels. This can include providing financial incentives, supporting research and development, and creating policies that encourage feedstock diversification and sustainable practices.

In conclusion, while feedstock availability and cost pose significant challenges to the biofuels industry, there are various strategies that can be employed to overcome these hurdles. By diversifying feedstock sources, embracing technological advancements, promoting sustainable farming practices, and fostering collaborations, we can pave the way for a more sustainable and accessible biofuels market. With continued efforts and advancements, biofuels have the potential to revolutionize the energy landscape and contribute to a greener future for everyone.

Scaling Up Biodiesel Production for Commercial Use

Biodiesel has emerged as a promising alternative to traditional fossil fuels, offering a cleaner and more sustainable source of energy. As the world faces increasing concerns over climate change and the need to reduce greenhouse gas emissions, biodiesel presents an opportunity to transition toward a greener future. However, to fully harness the benefits of this renewable energy source, it is crucial to scale up biodiesel production for commercial use.

Scaling up biodiesel production involves expanding the manufacturing capacity to meet the growing demand for this biofuel. This process requires careful planning, investment, and technological advancements. One of the key challenges in scaling up biodiesel production is ensuring a sustainable and reliable feedstock supply. Traditional feedstocks such as soybean oil and rapeseed oil have limitations in terms of availability and cost-effectiveness. Therefore, researchers and industry experts are exploring alternative feedstocks, including algae, waste cooking oil, and animal fats, to ensure a continuous and abundant supply of raw materials.

In addition to feedstock availability, the development of efficient and cost-effective production technologies is essential for scaling up biodiesel production. Traditional methods such as transesterification have been widely used, but they often involve high energy consumption and produce significant amounts of waste. To address these challenges, innovative technologies such as enzymatic transesterification, supercritical fluid processing, and microwave-assisted transesterification are being explored. These advancements not only enhance the efficiency of biodiesel production but also reduce the environmental impact of the manufacturing process.

Another important aspect of scaling up biodiesel production is ensuring the quality and compatibility of the fuel with existing infrastructure. Biodiesel must meet certain standards and specifications to be used in conventional diesel engines without modifications. To achieve this, rigorous testing and quality control procedures are necessary throughout the production process. Additionally, collaborations between biodiesel producers, engine manufacturers, and regulatory bodies are vital to establish industry standards and create a seamless transition to widespread biodiesel use.

Scaling up biodiesel production for commercial use requires a collective effort from various stakeholders, including policymakers, investors, researchers, and industry leaders. Governments can play a crucial role by providing incentives and supportive policies to encourage investment in biodiesel production facilities. Financial institutions can offer funding options to facilitate the expansion of biodiesel manufacturing. Researchers and innovators can continue to explore new feedstocks and production technologies to improve efficiency and reduce costs. Lastly, consumers can contribute by choosing biodiesel as a clean and renewable alternative to traditional fuels.

In conclusion, scaling up biodiesel production is a necessary step towards achieving a sustainable energy future. By addressing challenges related to feedstock availability, production technologies, and fuel quality, we can unlock the full potential of biodiesel for commercial use. With global efforts and collaborations, biodiesel can make a significant contribution to reducing greenhouse gas emissions and mitigating the impacts of climate change.

Exploration of Alternative Production Techniques

In recent years, the demand for renewable energy sources has grown significantly due to increasing concerns about climate change and the need to reduce greenhouse gas emissions. One such alternative that has gained considerable attention is biodiesel, a renewable fuel made from vegetable oils or animal fats. Biodiesel offers a promising solution to reduce our dependence on fossil fuels and mitigate the environmental impact of traditional transportation fuels.

This chapter delves into the exploration of alternative production techniques for biodiesel, aiming to provide a comprehensive understanding of the advancements and innovations in this field. As the demand for biodiesel continues to rise, it becomes essential to develop more efficient and sustainable production methods.

The use of traditional feedstocks such as soybean oil and rapeseed oil has its limitations, including competition with food crops and potential land-use conflicts. Therefore, researchers have been exploring alternative feedstocks that can be grown on marginal lands without competing with food production. These include non-food oil crops like jatropha, camelina, and algae, which have shown immense potential for biodiesel production.

Additionally, the development of advanced production techniques, such as enzymatic and microbial processes, has revolutionized the biodiesel industry. These alternative techniques offer several advantages, including higher conversion rates, milder reaction conditions, and reduced energy consumption. Enzymatic processes, for instance, utilize specific enzymes that break down the triglycerides present in feedstock into glycerol and fatty acid methyl esters, which are the main components of biodiesel. Microbial processes involve the use of microorganisms such as bacteria or yeast to convert feedstock into biodiesel through fermentation or other metabolic pathways.

Furthermore, the integration of biorefineries with biodiesel production has gained attention as a means to enhance sustainability and economic viability. Biorefineries enable the production of multiple products from a single feedstock, allowing for the utilization of by-products and waste streams, thereby reducing waste and maximizing resource efficiency.

The exploration of alternative production techniques for biodiesel is an ongoing endeavor, with researchers continuously striving to improve efficiency, reduce costs, and enhance sustainability. This chapter aims to provide readers with valuable insights into the latest breakthroughs and innovations in the biodiesel industry, encouraging the adoption of renewable energy sources for a cleaner and greener future.

Whether you are a biofuels enthusiast, a researcher, a student, or simply curious about renewable energy, this chapter will deepen your knowledge and appreciation for the advancements in biodiesel production techniques. Stay tuned to discover how these innovations are shaping the future of renewable energy and contributing to a more sustainable planet.

Chapter 8: Case Studies: Successful Biodiesel Projects

Biodiesel Production in Country X: Lessons Learned and Achievements

Introduction:

In recent years, the global demand for renewable energy sources has prompted significant advancements in the field of biofuels. Among these, biodiesel has emerged as a promising alternative to traditional fossil fuels. This subchapter explores the experiences and accomplishments of biodiesel production in Country X, shedding light on the lessons learned and achievements that have propelled this nation forward in the realm of renewable energy.

Country X's Commitment to Renewable Energy: Country X recognized the importance of reducing its dependency on fossil fuels and took proactive measures to promote the production and utilization of biodiesel. By leveraging its abundant agricultural resources, the nation embraced the potential of biofuels and embarked on a journey towards a greener and more sustainable future.

Investment in Research and Development: One of the key lessons learned in Country X's biodiesel production journey is the significance of investing in research and development. By fostering a culture of innovation and collaboration, the nation succeeded in developing cutting-edge technologies and refining existing processes. This commitment to research has not only enhanced biodiesel production efficiency but also opened doors for new applications and advancements in the biofuels sector.

Partnerships and International Collaborations: Country X understood the importance of partnerships and international collaborations in accelerating the growth of biodiesel production. By forging alliances with other nations, research institutions, and industry experts, the country was able to access valuable knowledge, expertise, and resources. These collaborations played a crucial role in fast-tracking technological advancements and knowledge sharing, driving the overall progress of the biofuels industry.

Sustainable Feedstock Production: Another significant achievement in Country X's biodiesel production is the establishment of sustainable feedstock production systems. The nation carefully balanced the need for renewable energy with environmental conservation, ensuring that the feedstock for biodiesel production was sourced responsibly. By implementing sustainable agricultural practices and promoting the cultivation of biodiesel-specific crops, the country achieved a delicate harmony between energy security and ecological preservation.

Economic and Social Impacts: Country X's biodiesel production endeavors have not only contributed to a cleaner environment but also yielded positive economic and social impacts. The industry has created employment opportunities, particularly in rural areas, stimulating economic growth and reducing poverty. Moreover, the increased use of biodiesel has helped mitigate greenhouse gas emissions, improving air quality and public health in the process.

Conclusion:
The biodiesel production journey in Country X serves as an inspiration for nations worldwide. By embracing innovation, fostering partnerships, and prioritizing sustainability, the country has unlocked

the potential of biofuels and achieved remarkable progress in renewable energy. The lessons learned and achievements made in Country X provide valuable insights for policymakers, researchers, and industry professionals seeking to advance the biofuels sector and build a more sustainable future.

Biodiesel Application in the Transportation Sector: Case Study Y

Introduction

In recent years, the transportation sector has witnessed a growing interest in alternative fuels due to concerns over environmental sustainability and the depletion of fossil fuel resources. Biodiesel, a renewable energy source made from organic materials such as vegetable oils and animal fats, has emerged as a viable and eco-friendly alternative to traditional diesel fuel. This subchapter explores a case study, Case Study Y, to highlight the successful application of biodiesel in the transportation sector.

Case Study Y: Implementing Biodiesel for a Greener Fleet

Case Study Y focuses on a transportation company that operates a large fleet of vehicles, including trucks and buses. In an effort to reduce their carbon footprint and comply with environmental regulations, the company decided to explore the use of biodiesel as a fuel source. They collaborated with local biofuel producers and conducted extensive research to ensure the feasibility and compatibility of biodiesel with their existing fleet.

Benefits of Biodiesel in Transportation

The case study highlights the numerous benefits of biodiesel in the transportation sector. Firstly, biodiesel significantly reduces greenhouse gas emissions, including carbon dioxide and particulate matter, compared to conventional diesel fuel. This reduction in emissions contributes to improved air quality and reduced health risks for both drivers and the community. Secondly, biodiesel is a renewable energy source, unlike fossil fuels, which are finite resources. By utilizing biodiesel, the company actively contributes to the transition towards a sustainable energy future.

Challenges and Solutions

The implementation of biodiesel in the transportation sector is not without challenges. The case study outlines the potential issues faced by the company, such as engine compatibility, cold weather performance, and fuel availability. However, the company successfully overcame these challenges through a combination of engine modifications, strategic fuel blending, and establishing partnerships with local biofuel producers to ensure a consistent supply of biodiesel.

Conclusion

Case Study Y serves as an exemplary illustration of the successful application of biodiesel in the transportation sector. By adopting biodiesel, the company was able to reduce emissions, promote sustainability, and enhance their public image as an environmentally conscious organization. This case study demonstrates the potential of biodiesel as a viable and eco-friendly alternative to conventional diesel fuel, making it an attractive option for other transportation companies and individuals seeking to make a positive impact on the environment.

Overall, biodiesel offers a promising solution to the environmental challenges faced by the transportation sector. As the world continues to seek sustainable energy alternatives, biodiesel breakthroughs, such as the one highlighted in Case Study Y, will play a pivotal role in shaping a greener and more sustainable future for all.

Biodiesel as a Community-Based Renewable Energy Solution: Case Study Z

In today's world, the need for renewable energy solutions has become increasingly evident. Fossil fuels, the primary source of energy for decades, are not only finite but also contribute significantly to environmental degradation and climate change. As a result, communities worldwide are exploring alternative energy sources, such as biodiesel. This subchapter aims to present a compelling case study, showcasing the successful implementation of biodiesel as a community-based renewable energy solution.

Case Study Z revolves around a small town that decided to take charge of its energy future by embracing biodiesel production. The town's commitment to sustainability and reducing its carbon footprint drove the initiative, which quickly gained traction among its residents. The project began by establishing partnerships with local farmers, who were encouraged to grow oilseed crops, such as soybeans or canola, to be used as feedstock for biodiesel production.

The community actively participated in every step of the process, from growing the crops to operating the biodiesel conversion facility. This involvement not only ensured a reliable supply of feedstock but also created local jobs and economic opportunities. The town's residents also actively used the biodiesel produced, further reducing their dependence on fossil fuels and supporting the local economy.

The success of Case Study Z lies in its ability to address various challenges associated with biodiesel production and consumption. The town implemented robust quality control measures to ensure the production of high-quality biodiesel that met industry standards. Additionally, the community actively engaged in public awareness campaigns to educate residents about the benefits of biodiesel and encourage its use.

Furthermore, Case Study Z highlights the importance of collaboration and knowledge sharing between communities. The town actively sought partnerships with other neighboring communities to exchange experiences, best practices, and lessons learned. By doing so, they were able to overcome obstacles more effectively and accelerate the adoption of biodiesel as a renewable energy solution in the region.

The success of Case Study Z serves as an inspiration and blueprint for other communities interested in implementing their renewable energy solutions. By embracing biodiesel, communities can reduce their dependence on fossil fuels, create local jobs, stimulate economic growth, and contribute to a sustainable future. This case study demonstrates that community-based initiatives can make a significant impact and act as a catalyst for broader change.

In conclusion, the subchapter "Biodiesel as a Community-Based Renewable Energy Solution: Case Study Z" presents a real-life example of how a small town successfully implemented biodiesel production and consumption. It highlights the benefits of community involvement, partnerships with local farmers, quality control measures, public awareness campaigns, and collaboration between communities. By sharing this case study, the book aims to inspire and empower readers to explore biodiesel as a viable renewable energy option in their own communities.

Chapter 9: Policy and Regulatory Framework for Biodiesel

Government Incentives and Support for Biodiesel Production

In recent years, there has been a growing recognition of the need to reduce our dependency on fossil fuels and transition towards sustainable energy sources. Biodiesel, a renewable fuel made from organic materials such as vegetable oils and animal fats, has emerged as a promising alternative to traditional diesel fuel. To encourage the production and use of biodiesel, governments around the world have implemented various incentives and support measures.

One of the key incentives provided by governments is financial support in the form of grants, loans, and tax incentives. These financial incentives help biodiesel producers offset the higher production costs associated with renewable fuels. They can also provide funding for research and development activities, helping to drive innovation in the industry. Additionally, governments often offer tax credits to consumers who use biodiesel, making it a more attractive option for individuals and businesses alike.

Furthermore, governments have established regulatory frameworks and standards to ensure the quality and sustainability of biodiesel production. These regulations help create a level playing field for producers and ensure that biodiesel meets the necessary performance and environmental standards. By setting clear guidelines, governments give confidence to investors, thereby encouraging more investment in biodiesel production facilities.

Many governments are also actively promoting biodiesel through public awareness campaigns and educational initiatives. These efforts aim to inform and educate the public about the benefits of biodiesel, such as reduced greenhouse gas emissions and improved air quality. By increasing public knowledge and understanding, governments hope to create a demand for biodiesel, further stimulating its production and use.

In addition to financial incentives and awareness campaigns, governments often establish research and development programs to support the advancement of biodiesel technology. These programs fund scientific research and collaboration between industry and academia, leading to breakthroughs and innovations in biodiesel production processes. By investing in research, governments demonstrate their commitment to renewable energy and help drive the industry forward.

Overall, government incentives and support play a crucial role in promoting and advancing biodiesel production. By providing financial assistance, establishing regulatory frameworks, raising public awareness, and investing in research and development, governments around the world are driving the growth of the biodiesel industry. With continued support, biodiesel has the potential to become a significant player in the global energy landscape, contributing to a more sustainable and environmentally friendly future for all.

International Standards and Certification for Biodiesel

In today's globalized world, the importance of establishing international standards and certifications for biodiesel cannot be overstated. As the demand for renewable energy sources continues to soar, biodiesel has emerged as a viable alternative to fossil fuels. To ensure its widespread adoption and facilitate a seamless global market, it is crucial to have a set of uniform standards and certifications that guarantee the quality and sustainability of biodiesel production.

The development and implementation of international standards for biodiesel are primarily driven by organizations like the International Organization for Standardization (ISO) and the American Society for Testing and Materials (ASTM). These standards cover various aspects of biodiesel production, including feedstock requirements, production processes, fuel quality, and environmental sustainability.

One of the key standards is the ASTM D6751, which specifies the requirements for biodiesel fuel blendstock. This standard ensures that biodiesel meets certain purity and performance criteria, making it compatible with existing diesel engines and infrastructure. By adhering to this standard, biodiesel producers can guarantee the quality and consistency of their product, instilling confidence in consumers and promoting market growth.

In addition to fuel quality, sustainability is another crucial aspect of biodiesel production. The Roundtable on Sustainable Biomaterials (RSB) has developed a comprehensive certification system that assesses the environmental, social, and economic sustainability of biofuels. This certification provides assurance that the biodiesel has been produced using sustainable practices, minimizing its impact on ecosystems and communities.

International standards and certifications play a vital role in facilitating trade and ensuring fair competition in the biodiesel market. By establishing a level playing field, these standards promote transparency and trust among producers, consumers, and policymakers. They also enable cross-border trade by harmonizing technical requirements and eliminating trade barriers.

For consumers, international standards and certifications provide a means to make informed choices and support sustainable biofuels. By looking for certified biodiesel products, individuals can contribute to reducing greenhouse gas emissions, improving air quality, and supporting rural development.

In conclusion, international standards and certifications are essential for the growth and sustainability of the biodiesel industry. They provide a framework for ensuring fuel quality, promoting sustainability, and facilitating global trade. By adhering to these standards, biodiesel producers and consumers can be confident in the environmental and social benefits of this renewable energy source. As the world transitions towards a greener future, international standards and certifications will continue to play a crucial role in driving innovation and advancing renewable energy.

Environmental Regulations and Sustainability Criteria

In recent years, the global community has witnessed a growing concern for the environment and the urgent need to find sustainable alternatives to fossil fuels. As a result, the biofuels industry has gained significant attention as a potential solution to mitigate climate change and reduce dependence on non-renewable energy sources. This subchapter explores the crucial role of environmental regulations and sustainability criteria in the biofuels sector.

Environmental regulations play a vital role in ensuring that the production and utilization of biofuels are environmentally friendly and do not cause further harm to the planet. Various countries and international organizations have established stringent guidelines to monitor and regulate the biofuels industry. These regulations aim to minimize greenhouse gas emissions, protect biodiversity, and promote sustainable agricultural practices.

One of the key aspects of environmental regulations is the establishment of sustainability criteria. These criteria define the minimum requirements that biofuels must meet to be considered environmentally sustainable. They typically include factors such as greenhouse gas emissions, land use change, water consumption, and protection of natural habitats. By defining and enforcing these criteria, regulators ensure that biofuels contribute positively to the environment and do not exacerbate existing environmental issues.

Sustainability criteria also address social aspects, such as human rights, labor conditions, and community development. This holistic approach ensures that the biofuels industry not only focuses on environmental sustainability but also promotes social and economic well-being. By adhering to these criteria, biofuel producers contribute to the overall sustainable development goals of communities and countries.

Furthermore, environmental regulations and sustainability criteria serve as a framework for certification and labeling schemes. These schemes provide consumers with the necessary information to make informed choices about the biofuels they use. By opting for certified biofuels that meet strict sustainability criteria, individuals and organizations can actively contribute to a greener and more sustainable future.

It is essential for stakeholders in the biofuels industry, including producers, policymakers, and consumers, to understand and comply with environmental regulations and sustainability criteria. By doing so, they can ensure the long-term viability and success of the biofuels sector while mitigating the negative impacts of traditional fossil fuels on the environment.

In conclusion, environmental regulations and sustainability criteria play a crucial role in the biofuels industry. They establish guidelines and minimum requirements to ensure that biofuels are produced and utilized in an environmentally sustainable manner. By adhering to these regulations, stakeholders can contribute to a greener future and reduce the world's dependence on non-renewable energy sources.

Chapter 10: Conclusion and Future Outlook

Summary of Key Biodiesel Breakthroughs

Biodiesel Breakthroughs: Innovations and Advances in Renewable Energy presents a comprehensive overview of the latest breakthroughs in the field of biodiesel production and utilization. This subchapter, titled "Summary of Key Biodiesel Breakthroughs," aims to provide a concise yet informative summary of the most significant advancements in the biofuel industry. The content is tailored for a diverse audience, including experts, enthusiasts, and individuals interested in renewable energy sources.

The biodiesel industry has witnessed remarkable progress in recent years, with numerous breakthroughs enhancing the efficiency, sustainability, and cost-effectiveness of biofuel production. This subchapter highlights several key innovations, beginning with the development of advanced feedstocks. Researchers have successfully explored non-food crops such as algae, jatropha, and camelina as potential sources of biodiesel, reducing competition with food supply and increasing overall sustainability.

Another major breakthrough is the improvement in conversion technologies. Scientists have developed more efficient and environmentally friendly methods for converting feedstocks into biodiesel, such as enzymatic transesterification and supercritical methanol processes. These advancements have significantly reduced energy consumption, waste generation, and overall production costs.

Furthermore, the integration of waste streams into biodiesel production has been a game-changer. By utilizing waste materials like animal fats, used cooking oil, and even sewage, researchers have not only reduced waste disposal problems but also increased the availability of feedstocks, contributing to a more sustainable biofuel industry.

Additionally, researchers have made substantial progress in optimizing the performance of biodiesel in engines. Through the development of novel additives and blending techniques, biodiesel can now be used in conventional diesel engines without significant engine modifications. This breakthrough has opened up new opportunities for widespread adoption of biodiesel as a viable alternative to fossil fuels.

Lastly, policy and market developments have played a crucial role in advancing the biodiesel industry. Government incentives, such as tax credits and renewable fuel standards, have encouraged investment in biofuel production, leading to increased research and development efforts. Additionally, the growing demand for sustainable and environmentally friendly fuel options has spurred market growth and created new opportunities for biofuel producers.

In conclusion, this subchapter provides a concise summary of key biodiesel breakthroughs, highlighting advancements in feedstocks, conversion technologies, waste utilization, engine performance, and policy support. These breakthroughs are driving the biofuel industry towards a more sustainable and economically viable future.

The Role of Biodiesel in Achieving Renewable Energy Targets

In recent years, the global focus on renewable energy has grown exponentially, as the need for sustainable alternatives to fossil fuels becomes increasingly urgent. Biodiesel, a clean-burning alternative fuel derived from renewable sources such as vegetable oils and animal fats, has emerged as a promising solution to meet these renewable energy targets. In this subchapter, we will explore the significant role that biodiesel plays in achieving these goals.

First and foremost, biodiesel offers a viable and sustainable alternative to traditional fossil fuels, such as gasoline and diesel. Unlike conventional fuels, biodiesel reduces greenhouse gas emissions, mitigating the impact of climate change and improving air quality. Its production and combustion result in significantly lower emissions of carbon dioxide, particulate matter, and sulfur oxides. As a result, biodiesel plays a crucial role in reducing the carbon footprint of transportation, which is a major contributor to global emissions.

Furthermore, biodiesel is a versatile fuel that can be seamlessly integrated into existing infrastructure. It can be used in conventional diesel engines without requiring any significant modifications, making it an accessible and practical choice for consumers and industries. This compatibility allows for a smooth transition from fossil fuel consumption to a more sustainable energy source, without the need for extensive infrastructure overhauls.

Moreover, the production of biodiesel has numerous economic and environmental benefits. The cultivation of feedstock crops for biodiesel production provides opportunities for rural development and agricultural diversification. Additionally, the production process generates valuable by-products such as glycerin, which can be used in various industries. By utilizing these by-products, biodiesel production contributes to a more circular economy and reduces waste.

In terms of energy security, biodiesel offers a viable alternative to fossil fuel imports, reducing dependence on volatile global oil markets. By promoting domestic production and utilizing locally available resources, countries can enhance their energy independence and reduce the risks associated with oil price fluctuations and geopolitical tensions.

In conclusion, biodiesel plays a pivotal role in achieving renewable energy targets by providing a clean, sustainable, and versatile fuel option. Its ability to reduce greenhouse gas emissions, compatibility with existing infrastructure, economic benefits, and contribution to energy security make it a crucial component of the transition to a more sustainable and resilient energy future. By embracing biodiesel as a viable alternative, we can collectively work towards a greener and more sustainable planet for future generations.

Future Prospects and Emerging Trends in Biodiesel Production

Biodiesel production has witnessed significant advancements and innovations in recent years, paving the way for a promising future in renewable energy. As the world faces the challenges of climate change and the depletion of fossil fuels, the demand for sustainable energy sources, such as biodiesel, continues to grow. This subchapter explores the future prospects and emerging trends in biodiesel production, highlighting the potential benefits and advancements that lie ahead.

One of the key future prospects in biodiesel production is the development of advanced feedstocks. Traditionally, biodiesel has been produced from edible crops such as soybean and rapeseed. However, the use of non-edible feedstocks, such as algae, jatropha, and waste cooking oil, is gaining momentum. These feedstocks offer several advantages, including higher oil yields, reduced competition with food crops, and the potential to utilize waste materials effectively. The research and development in this area hold great promise for expanding the biodiesel industry and reducing its environmental impact.

Another emerging trend in biodiesel production is the integration of biotechnology and genetic engineering. Scientists are exploring the modification of microorganisms to enhance their oil-producing capabilities, resulting in higher biodiesel yields. This approach not only increases efficiency but also reduces the reliance on energy-intensive processes involved in traditional biodiesel production. Furthermore, the genetic modification of crops can improve their oil content, making them ideal feedstocks for biodiesel production. These advancements in biotechnology have the potential to revolutionize the biodiesel industry and make it more sustainable and economically viable.

Additionally, the future of biodiesel production involves the optimization of production processes. Researchers are continuously working on improving the efficiency of transesterification, the chemical reaction that converts vegetable oils or animal fats into biodiesel. This includes the development of novel catalysts, such as enzymes and heterogeneous catalysts, which can enhance the reaction rate and reduce energy consumption. Furthermore, the utilization of innovative reactor designs and process intensification techniques can lead to higher biodiesel yields and reduced production costs.

In conclusion, the future prospects and emerging trends in biodiesel production are promising and offer a sustainable solution to the world's energy needs. The development of advanced feedstocks, integration of biotechnology, and optimization of production processes are key areas that hold significant potential. As these innovations continue to evolve, biodiesel production will become more efficient, economically viable, and environmentally friendly. By embracing these advancements, we can pave the way for a greener and more sustainable future for everyone, while addressing the niche of biofuels.